Shape and size

Study the shapes on the surfaces of houses, schools, churches, museums, and factories, and see how many shapes you can find. Some modern buildings have flat roofs and are oblong in appearance, but many have a variety of shapes. The way a building looks can often tell you something about its construction.

The photo above right shows a building which was built about four hundred years ago. How many different shapes can you find at roof level? At ground level? On the patterned walls? Describe the shapes. Find out how the materials used influenced these shapes.

A *geoboard* is a useful device for studying geometry. Shapes can be made with coloured elastic bands. Two geoboards are shown above right. In one, the pins are arranged in a square pattern, and in the other in a triangular pattern.

A lattice is a system of lines that cross each other. The pins on a geoboard represent the points on a lattice where the lines cross. The photographs below show how to make geoboards from lattices.

Use elastic bands to investigate the possibilities of your geoboards. How many shapes can you make on each? What effect does the difference in the arrangement of pins have?

Old houses in Chester, UK. Geometric shapes can be seen on new as well as old buildings

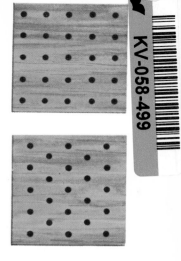

Geoboards with pins arranged in different patterns

Dot lattice paper and coloured pencils can be used to record situations made with the coloured elastic bands on the geoboard.

Exercise 1
Two geoboards are shown on the right. On each board is a red elastic band joining two pins. These red bands represent *line segments* of the same length. Use elastic bands to reproduce each of the shapes below on geoboards of the same size. Each colour represents a particular length. Which shapes can be made on
a. the square lattice geoboard?
b. the triangular lattice geoboard?

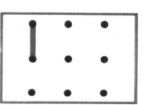
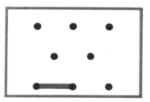

How to make a geoboard

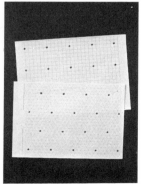

For the lattices you need:
A piece of grid paper 30 cm by 30 cm of each of these patterns marked at centimetre intervals, a hard pencil, a ruler

On the square lattice, mark a large dot at every sixth point as shown

On the triangular lattice, mark a large dot at every sixth point as shown

For each board you need:
A 30 cm by 30 cm piece of 9 mm plywood, about 30 $\frac{5}{8}$ inch escutcheon pins, a hammer

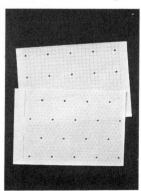

Pin the lattice over the board. Hammer in the pins at the points marked by the dots. Remove the lattice

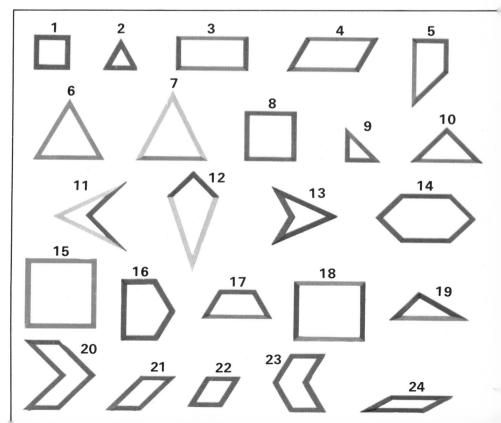

Naming points and angles

Exercise 2

1. In the first diagram, right, the positions of the points in blue are given by the pairs of numbers in brackets. What rules have been used to name these points?

2. Make a copy of the diagram. Keeping to the same rules, name each of the green points.

3. What name should be given to the red point?

4. A different geoboard is shown in the second diagram on the right. What rules have been used here to name the points? Copy the diagram.

5. Keeping to the rules of question 1, name the green points.

6. What name should be given to the red point?

7. Circles which have a common centre are called *concentric* circles. The third diagram on the right shows a way of naming the points on a geoboard whose pins are arranged in concentric circles. Copy the diagram. Name the green points.

8. Why does the red dot have the same name as the red dots on the other boards?

9. The diagram below shows a triangular lattice geoboard with the rules changed. Copy the picture. Write down the new names for the green points.

10. What name is given to the red point?

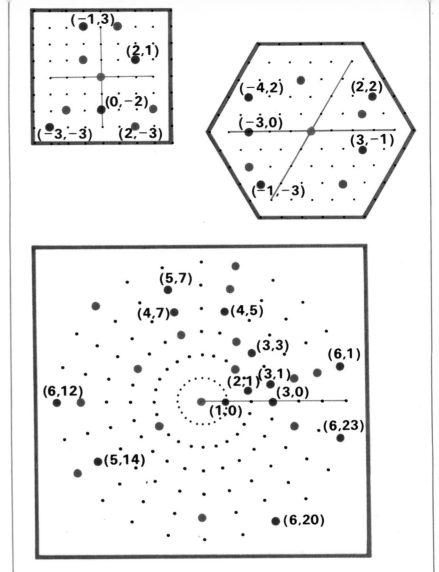

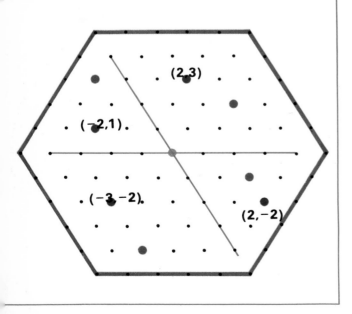

Ordered pairs

In each case, the name of a point is written as an *ordered pair* of numbers. The order of the numbers is important. For example, the point named as (3,2) is a different one from the point named (2,3). Such a system of names is called a *co-ordinate system*.

Any line used for reference when counting these numbers—the red lines in the diagrams above—is called an *axis* (plural *axes*). The red point is called the *origin* of the co-ordinate system, and is the point (0,0). This point has the same name in each of the systems above.

On the square lattice geoboard, right, a blue band represents the horizontal axis b, and a green band the vertical axis g. The position of a lattice point is given by the ordered pair (b,g). For example, the pink and purple lines meet at (1,2). The intersection of b and g is the origin (0,0).

The point (1,2) is named by referring it to the blue and green lines

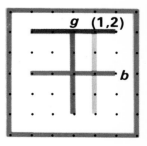

Exercise 3
Use blue and green bands to represent axes on your square lattice geoboard.
1. Put a yellow band round each set of pins described by these ordered pairs:

$A = \{(2,-2), (2,-1), (2,0), (2,1), (2,2)\}$
$B = \{(-1,-2), (-1,-1), (-1,0), (-1,1), (-1,2)\}$

a. What do you notice about the first number of each pair in each set?
b. Compare the pairs in A and B. What do you notice about the second numbers?

2. Put a red elastic band round each set of pins described by these ordered pairs:

$C = \{(-1,-2), (0,-2), (1,-2), (2,-2)\}$
$D = \{(-1,2), (0,2), (1,2), (2,2)\}$

a. What do you notice about the second number of each pair in each set?
b. Compare the pairs in C and D. What do you notice about the first numbers?

3. A and B have number pairs which are the same as those in C and D. Explain this relation in terms of the red and yellow bands and the pins.
4. How many pins enclosed by each red band lie between the yellow bands?
5. How many pins enclosed by each yellow band lie between the red bands?
6. Lines which are the same distance apart are *parallel* to each other. How do your investigations help to show that:
a. the yellow bands are parallel?
b. the red bands are parallel?

Repeat Exercise 3 on a triangular lattice geoboard. Count the number of points between the red lines along each line of the lattice shown below in yellow. There are the same number of points in each case, and so the lines are parallel.

Parallel lines as represented on a square or triangular lattice can be defined in terms of 'points between'. The next diagram shows that this definition does not hold for a geoboard with pins arranged in concentric circles. Why? Instead of 'points between' we shall use angles.

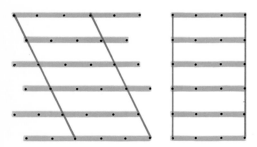

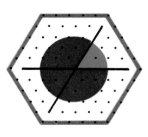

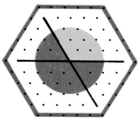

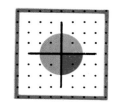

Using a protractor to measure an angle

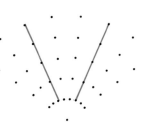

Angles are the best way of testing whether lines are parallel

Angle sets
On each of the three diagrams on the left, a circle has been drawn round the origin. Copy the diagrams and cut out the circles. Cut out the coloured parts. They are called *sectors*.

In the diagram at the bottom of the page, the blue sector has been placed on a *circular protractor*. This is an instrument for measuring the size of angles. It has 360 equal divisions called *degrees*, and usually written as 360°. The outer scale is used for measuring clockwise, the inner for measuring anti-clockwise, as shown here.

Measure the parts you have cut out. Now study these angle definitions before doing the exercise. An angle of 90° is called a *right* angle. An angle greater than 0° and less than 90° is called an *acute* angle. An angle greater than 90° and less than 180° is called an *obtuse* angle. An angle greater than 180° and less than 360° is called a *reflex* angle.

Exercise 4
Copy and complete the following table.

Colour	Angle in degrees	Angle as fraction of circle	Right angle	Acute	Obtuse	Reflex
green	90°	$\frac{90}{360} = \frac{1}{4}$	✓			
blue	60°					
yellow	120°					
pink	270°					
purple	300°					
red	240°					

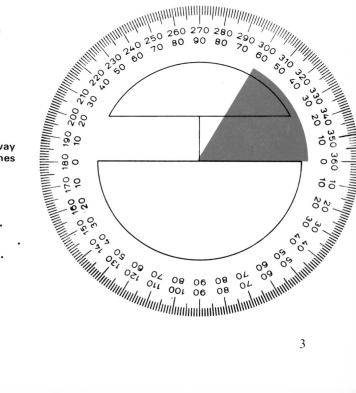

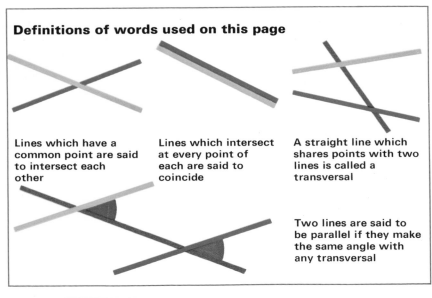

Definitions of words used on this page

Lines which have a common point are said to intersect each other

Lines which intersect at every point of each are said to coincide

A straight line which shares points with two lines is called a transversal

Two lines are said to be parallel if they make the same angle with any transversal

Similar triangles

Study the definitions on the left. Now look at the middle diagram, below left. Copy and cut out sectors to fit the angles shown in black. Use these sectors to check that all the sets of points on the blue lines are parallel.

The bottom diagram on the left shows two triangles. Draw them on a square dot lattice. Draw a line to cut across both triangles. This line is called a *transversal*. Cut out sectors and check that the sides of one triangle are parallel to the sides of the other, as shown in the diagram. Repeat for different transversals.

Cut out another triangle which will fit exactly on top of the small one. Place this small triangle on top of the large one, and slide it along to see if it fits each of the corners in turn. Are the angles the same?

Triangles which have the same angles are called *similar triangles*. They are the same shape.

On your triangular lattice geoboard find more triangles with parallel sides. Draw these on a dot lattice, and test the angles to see whether the triangles are similar. Can you find similar triangles whose sides are not parallel?

Make another copy of the yellow triangles on a square dot lattice. Take your cut-out triangle which fits the small one and place it on one corner of the large one. Draw in its third side, as shown below, and put in the coloured angles. How many small triangles fit the large one? Draw them in, marking the colour of each angle.

The three angles of the small triangle meet together on the sides of the large one. What fraction of the space around each point do the red, brown, and green angles fill? What do these angles total in degrees?

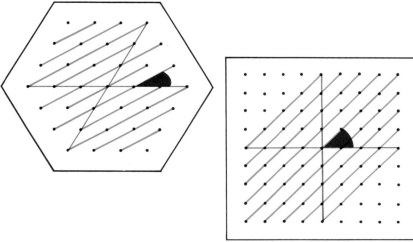

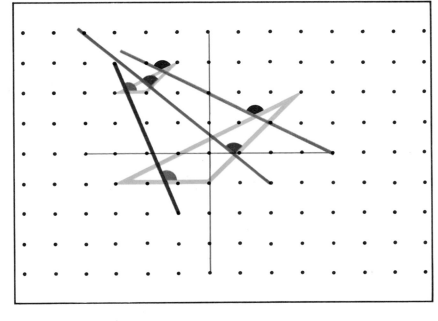

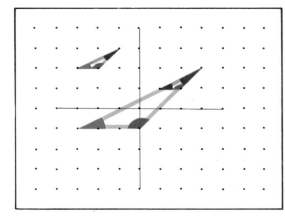

The elastic ruler

A simple but useful tool in geometry is an elastic ruler. To make elastic rulers you need a short length of shirring elastic, a piece of square-lattice paper, and a fine felt-tip pen.

Work in pairs, one to hold, the other to mark. Cut a piece of elastic about as long as this page. Put the ends together and cut in half. Stretch one half of the elastic along a line of dots on a square lattice. Mark the dots carefully on the elastic. Use the other half of elastic to make the second ruler.

Use your elastic ruler to divide lines into 2, 3, 4, 5, 6, 7 or more equal parts.

On the line below, the elastic ruler is used to mark seven equal spacings. The line is divided into two parts, one of 3 units, the other of 4 units. The relation of the first part to the second part can be written as the fraction $\frac{3}{4}$, or as the number pair 3:4. The relation is called a *ratio*. Notice that 6 to 8 divides fourteen equal spacings in the same ratio as 3 to 4 divided seven equal spacings. The ratios $\frac{3}{4}$ and $\frac{6}{8}$ are equal and the parts are said to be in *proportion*.

Copy the picture below onto square-dotted paper. The blue lines divide the purple line in the ratio 3 to 2. Draw four red lines through the red point to intersect the blue lines. Test with the elastic ruler that the blue lines divide each red line in the ratio 3 to 2. Draw red lines through other points to intersect the three blue lines, and test them with the elastic ruler. Repeat this exercise with differently spaced blue lines on sheets of dotted paper.

You have now discovered that three or more parallel lines divide all transversals in the same proportion. The diagrams at the bottom of the page show some ways in which this discovery can be put to use. Round, as well as flat, objects can be marked off in equal parts. Sometimes the elastic ruler is a more convenient method, sometimes the parallels.

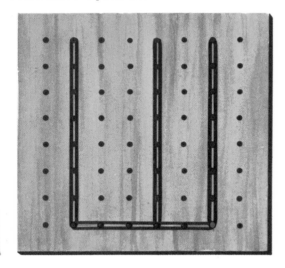

Using an elastic ruler to divide a line into different numbers of equal parts

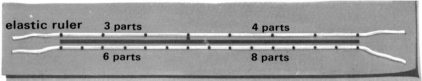

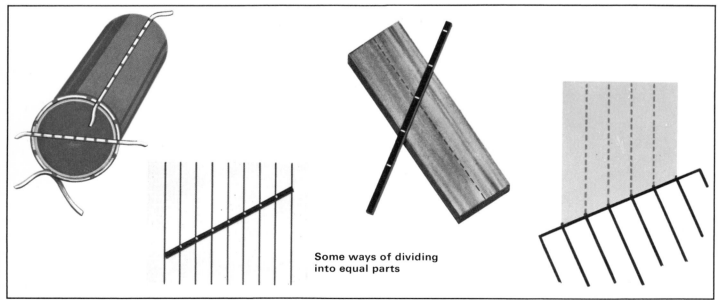

Some ways of dividing into equal parts

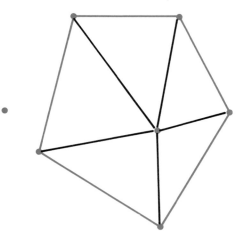

Dots and triangles

Take a sheet of paper, close your eyes and make six dots on it with a felt-tip pen. The result will probably look something like that above left. Join the dots with straight lines according to these rules:
1. The lines must cross only at a dot.
2. Every pair of dots must be joined, unless to do so would break rule 1.

The figure you get by obeying these rules is *closed*, which means that its boundary has no ends. The figure is divided into triangular regions made up from three dots only. These are called *three-dot triangles*.

In the diagram above right, the boundary is blue. How many boundary dots, middle dots, and triangles has your shape?

The boundary must enclose a space. Is it possible to get more or less dots on the boundary? Try different arrangements of the six dots. How many different arrangements can you make? Now find four different arrangements on a circle geoboard.

Complete the table below. What do you notice?

On your original drawing, put another dot *on* or *inside* the boundary. Make up any new three-dot triangles without breaking rule 1. How many extra triangles did you get? Note that at least one of the original triangles is no longer a three-dot triangle.

Boundary dots	Middle dots	Triangles
6		
5	1	5
4		
	3	

Exercise 5
1. Make three copies of the diagram on the left below. Put a dot:
a. on the boundary
b. within the boundary on a line
c. within the boundary in a space.
How many extra triangles do a, b, and c give?
2. A number of dots, below right, has been added to the first diagram. How many extra triangles are given by:
a. the blue dots?   b. the red dots?
c. Make up a rule that will give the number of triangles, if the number of boundary dots and middle dots are given.

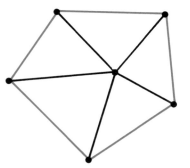 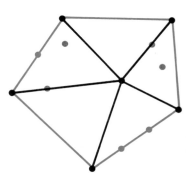

3. Make a copy of the diagram below. Put a point outside the boundary. Draw new lines.
a. Does the number of boundary dots **i** increase, **ii** stay the same, or **iii** decrease?
b. Make more copies of the diagram. Put a point outside the boundary in each, and draw new lines to show that **i**, **ii**, and **iii** are possible.
c. Do the examples in this question fit the rule you have made in question 2c?
d. Write out the rule in shorthand, using *B* for boundary, *M* for middle, and *T* for triangles.

The different way in which things connect is part of the branch of mathematics called *topology*. The rule you have found, $T = B + 2M - 2$, is a topological version of a theorem put forward by Pick in 1899.

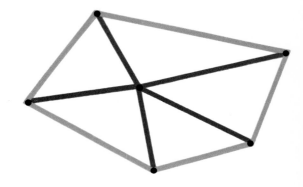

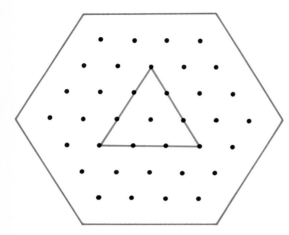

Three-dot triangles and lattices

Set up the diagram shown above on your triangle geoboard. Divide the large equilateral triangle into three-dot triangles in as many different ways as you can. Use different coloured elastic bands for triangles next to each other. Draw the pattern of triangles on a piece of lattice paper like that shown on the right.

Note that the solutions shown in the diagrams on the right are all considered to be the same. Each solution can be changed into any one of the others by turning it round or by looking at it in a mirror.

When you have found six different solutions, put them together to form a regular hexagon. Colour it so that no three-dot triangles next to each other share the same colour. There are 18 different ways.

Does each solution have nine three-dot triangles? Remember Pick's theorem.

The triangles are not all the same shape. Do they have the same area? The following exercise should help you to answer this question.

Exercise 6

1. Copy and complete the following table. The shapes are those used in the triangle patterns you made above.

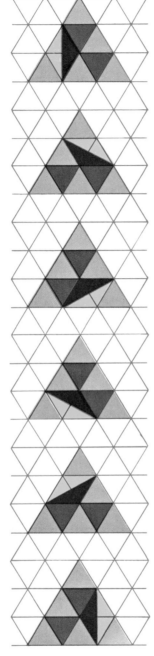

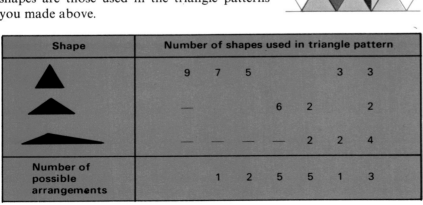

Shape	Number of shapes used in triangle pattern					
▲	9	7	5		3	3
▲ (flat)	—			6	2	2
▂ (long)	—	—	—	—	2	2	4
Number of possible arrangements	1	2	5	5	1	3

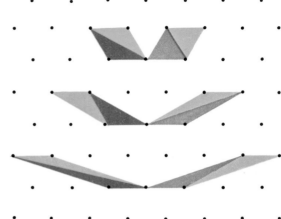

2. A figure with four sides and with opposite sides parallel is called a *parallelogram*. The diagonal halves the parallelogram. In the diagram above, the parallelograms are each divided by a diagonal into two three-dot triangles. Why are all these triangles equal in area?

The axes in the diagram below are marked in red. Write down the co-ordinates of the vertices of the blue triangle. The co-ordinates can also be written as the matrix $\begin{pmatrix} 2 & 4 & 1 \\ 0 & 1 & 3 \end{pmatrix}$.

Exercise 7

Here are the matrices for six triangles:

$$\begin{pmatrix} -5 & -2 & -1 \\ 1 & -1 & 1 \end{pmatrix} \quad \begin{pmatrix} 3 & 1 & 4 \\ -1 & -3 & -2 \end{pmatrix} \quad \begin{pmatrix} 0 & -1 & 2 \\ 5 & 2 & 4 \end{pmatrix}$$

$$\begin{pmatrix} 7 & 3 & 5 \\ 2 & 0 & -1 \end{pmatrix} \quad \begin{pmatrix} 4 & 5 & 3 \\ 5 & 3 & 2 \end{pmatrix} \quad \begin{pmatrix} -5 & -2 & -1 \\ 2 & 3 & 6 \end{pmatrix}$$

1. Draw the triangles on square-dotted paper.
2. Redraw them on triangle-dotted paper using axes at an acute angle.
3. Draw them again on triangle-dotted paper with axes at an obtuse angle.
4. Either by drawing three-dot triangles or by using Pick's theorem find their areas in three-dot triangles.

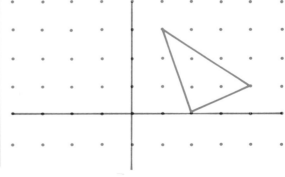

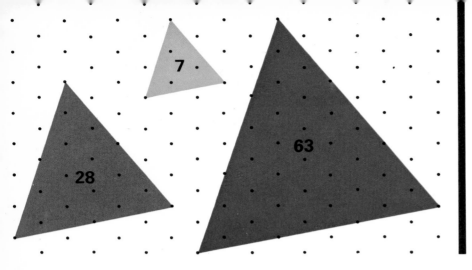
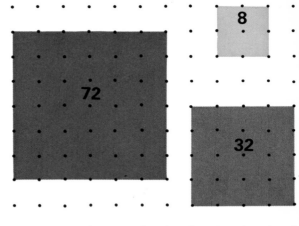

Similarity and area

Pick's theorem and the three-dot triangle unit of area provide an easy method of drawing many different shapes which have the same area.

Draw on square-dotted paper a whole sheet of different shapes, the area of each to be eleven three-dot triangles. Try to find examples which show all the different possible combinations of middle and boundary dots which give as much variety of shape as possible. Repeat on triangle-dotted paper.

Choose any two points on a sheet of triangle-dotted paper and join them. Find a third point so that all three points are equal distances apart. They form an *equilateral triangle*. Given any two points we can always find a third. Find the area of your figure in three-dot triangles.

On square-dotted paper, draw as many different-sized squares as you can. Using the three-dot triangle as a unit of area, find the area of each square. Draw as many equilateral triangles as you can on triangle-dotted paper and find their areas.

Use a number square like the one below to record your results. Draw squares in blue and triangles in red round any numbers which represent the areas of your shapes.

```
 1   2   3   4   5   6   7   8   9  10
11  12  13  14  15  16  17  18  19  20
21  22  23  24  25  26  27  28  29  30
31  32  33  34  35  36  37  38  39  40
41  42  43  44  45  46  47  48  49  50
51  52  53  54  55  56  57  58  59  60
61  62  63  64  65  66  67  68  69  70
71  72  73  74  75  76  77  78  79  80
81  82  83  84  85  86  87  88  89  90
91  92  93  94  95  96  97  98  99 100
```

The areas of the three triangles above are 7, 28, and 63 units. The areas of the three squares are 8, 32, and 72 units.

Exercise 8
1. Copy and complete the following table.

Area of triangle	Ratio	Side	Ratio
7 = 7 × 1	1		
28 = 7 × 4	4		
63 = 7 × 9	9		

2. Copy and complete the following table.

Area of square	Ratio	Side	Ratio
8 = 8 × 1			
32 = 8 × 4			4
72 =			

3. Complete the statement: The areas of similar figures are proportional to ... of the corresponding sides.

Exercise 9
1. A number multiplied by itself gives its square. For example, $5 \times 5 = 25$. 25 is the square of 5. We say that 5 is the *square root* of 25. The sign $\sqrt{}$ is used for square root, so $\sqrt{25}$ is 5. Fold a piece of triangle-dotted paper along a lattice line. Use this to measure $\sqrt{3}, \sqrt{7}, \sqrt{12}, \sqrt{28}, \sqrt{63}$. Find more accurate values from a table of square roots.

2. Repeat question 1 using a piece of square-lattice paper.

3. Complete this statement: On a triangle lattice the length of side of an equilateral triangle is ... of the number of three-dot triangles.

4. Complete this statement: On a square lattice, the length of side of a square is ... of ... the number of three-dot triangles.

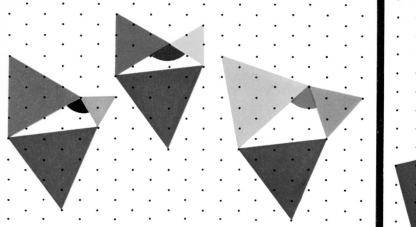
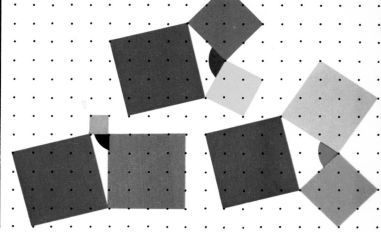

Theorem of Pythagoras

Use the green sector which you cut out earlier (see page 3) to test the black, purple, and blue sectors in the diagrams above. Which angle is a right angle? Which is an obtuse angle? Which is acute?

The area of each brown equilateral triangle in the diagrams above is 19 three-dot triangles. The area of each brown square is 34 three-dot triangles.

Exercise 10
1. Add together the number of three-dot triangles in the diagrams above in
a. the blue and red triangles
b. the orange and pink triangles
c. the yellow and green triangles.
2. Is the area of the brown triangle less than, greater than, or equal to the sum of the areas of
a. the blue and red triangles?
b. the orange and pink triangles?
c. the yellow and green triangles?
3. Add together the number of three-dot triangles in
a. the blue and red squares.
b. the orange and pink squares.
c. the yellow and green squares.

4. Is the area of the brown square less than, greater than, or equal to the sum of the areas of the:
a. blue and red squares?
b. orange and pink squares?
c. yellow and green squares?

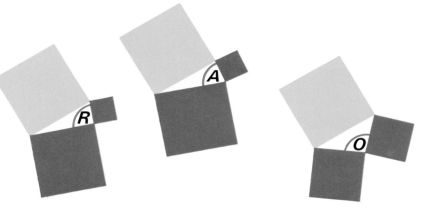

5. In the diagram above, R is a right angle, A is acute, and O is obtuse. For which of the angles R, A, and O is the area of the yellow square a. less than, b. greater than, c. equal to the sum of the areas of the green squares?

6. Copy the shapes in each set below. Arrange them so that they enclose a right-angled triangle. What is the relationship between the area of the shapes on the sides of each triangle?

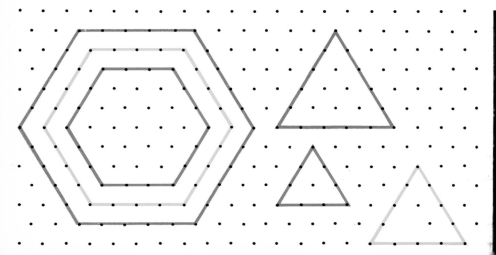
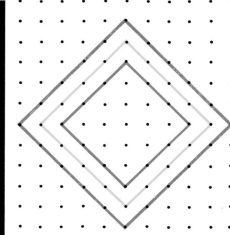

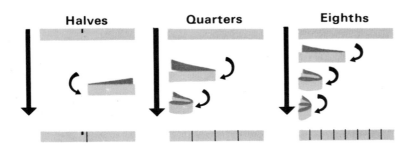

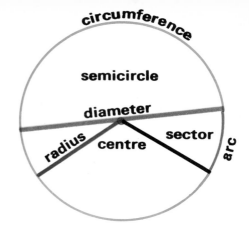

Paper-folding

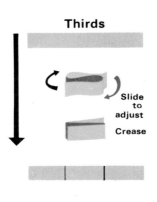

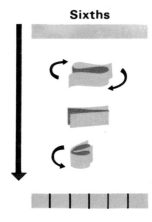

Dividing strips into equal parts
How good are you at guessing? Take a strip of paper and guess the middle of it. Mark it lightly in pencil. Now fold the strip to find the middle. Was your guess accurate?

Guess the middles of the two halves and again check your guesses by folding. Guess and check the middles of the four quarters. Practise with other strips until you are really good at guessing the middle.

Take another strip of paper. To divide it into three equal parts follow the steps shown in the diagram on the left. Fold one end in front and the other behind. Slide and roll the paper until the ends of the paper coincide with the folds. Use your nails to crease the folds.

To divide a strip into six equal parts, follow the steps in the second diagram on the left. Fold the strip into thirds, and then fold over to halve each of the thirds. Crease the folds.

To divide a strip into five equal parts, follow the steps in the diagram below. Fold it so that the extra at one end is half the length of the folded part. Check it by folding the folded part in two. Slide the paper to adjust the folds until the extra part is one quarter of the rest. Crease the folds. Halving folded fifths divides a strip into ten equal parts.

The method for seven equal parts is similar to that for fifths. Fold the paper so that the extra bit at one end is a third the length of the folded part. Check by folding the folded part into thirds. Slide the paper until these thirds equal the extra part. Crease the folds.

This basic method can be used to divide a strip into any odd number of equal parts. Halving an odd number of equal parts gives the basic method for any even number of equal parts.

Dividing circles to make polygons
An *arc* of a circle is any part of its boundary, or *circumference*, as it is called. In the diagram above, the circumference is shown in blue. The part shown in red is an arc. The diagram also shows the meanings of some other words.

Draw several circles with a compass and cut them out. Use a radius of about 6 cm. Many regular polygons can be made by folding paper circles. A polygon is any flat shape with straight sides. A *regular polygon* is one whose sides are all equal.

Equilateral triangle
The diagram below shows how to make an equilateral triangle from a circle. The dotted lines show the folds to be made next. Fold the circle in half along a diameter. Divide the circumference of the semicircle into three equal parts. Use the same method as for dividing the strip into three equal parts. Pinch only the edges. Open the circle out. The circumference is divided into six equal arcs.

Go over each of the six pinch marks with a pencil. Mark the centre of the circle. It is at the midpoint of the diameter. Fold every other pinch onto the centre. Tuck the last flap under the first. The resulting shape is an equilateral triangle.

Making an equilateral triangle from a circle

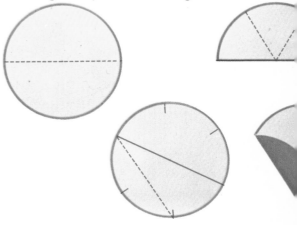

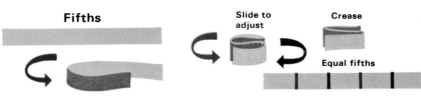

Making a hexagon from a circle

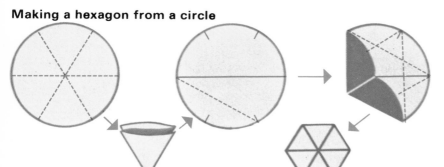

The Defense Department of the United States of America has its main offices in the Pentagon in Washington. The building is so called because it is a regular five-sided shape

Hexagon

Fold another circle in half and divide the circumference into three equal parts, as shown in the diagram above. Mark the centre and pinch marks in pencil. Fold each pinch mark onto the centre. The result is a hexagon.

These two shapes show the basic methods for making any regular polygon. Always start by folding a circle in half. For an odd number of sides, divide the semicircle into that number of equal parts. Open out the circle, and fold to the centre every other pinch mark. For an even number of sides, divide the semicircle into half that number of equal parts. Open out the circle, and fold to the centre every pinch mark.

Make several of each of the shapes in the diagram below. A pentagon has five equal sides, and can be called a 5-gon. Similarly, a hexagon could be called a 6-gon, a heptagon a 7-gon, an octagon an 8-gon, a nonagon a 9-gon, and so on.

A tidier way of folding away the unwanted paper is shown in the diagrams on the left. Fold the shape in the usual way, and then unfold it. Mark the lines shown here in red and blue. Fold on the red lines. Press in the blue creases on each side of the flap. Fold all the flaps down to complete the polygon.

Tessellations

How many equilateral triangles will fit together like tiles to fill all the space around a point? How many squares? How many regular hexagons? These patterns are called *regular tessellations*. They are made from uniform sets of regular shapes.

Can you make a regular tessellation with any other shapes on these pages? Explain why only certain regular shapes will form a regular tessellation.

Tiling patterns can be made using a combination of regular shapes. These are referred to as *semi-regular* tessellations. The sides of the square and the octagon in the diagram below are the same length. Trace and cut out these shapes. Use them to make a semi-regular tessellation.

The pattern on the front cover is made from equilateral triangles and squares. Make tessellations from the following combinations of regular shapes:
Hexagon and equilateral triangle.
Hexagon, square and equilateral triangle.
Duodecagon and equilateral triangle.

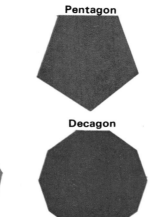

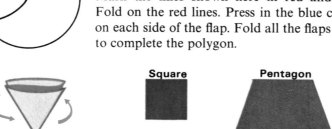

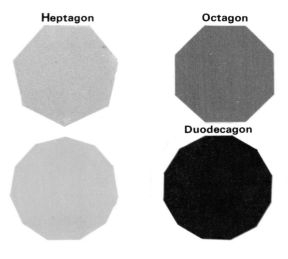

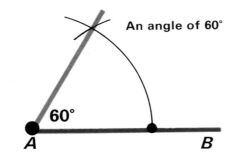
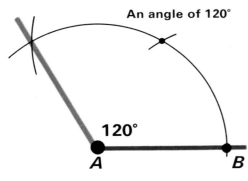
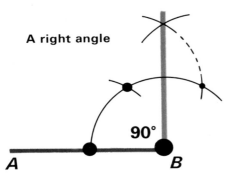

Some useful constructions

You will need only a straight edge, a pair of compasses, and a pencil for these constructions.

The three diagrams on the left show how to draw an angle of 60°, an angle of 120°, and a right angle. Start each time with the line labelled *AB*. The largest dot shows where to put the compass point first. The parts of the circle you draw give the next compass points. The dots showing the compass points in these diagrams get smaller as the construction progresses.

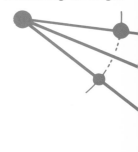

The diagrams above right show how to divide a line and an angle into two equal parts. Dividing into two equal parts is called *bisecting*. Why does the construction for bisecting a line give another way of constructing a right angle?

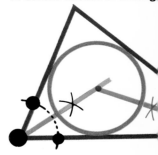

The second set of diagrams on the right show how to draw a circle inside and outside a triangle. For the circle inside first bisect two of the angles of the triangle. For the circle outside first bisect two of the sides.

The six diagrams below show how to draw an equilateral triangle, a square, a hexagon, an octagon, a pentagon, and a heptagon. Start each time from the purple line. Again the size of the compass dot gets smaller as the construction progresses. Notice that the construction for 60° is used for the equilateral triangle, and the construction for 90° is used for the square.

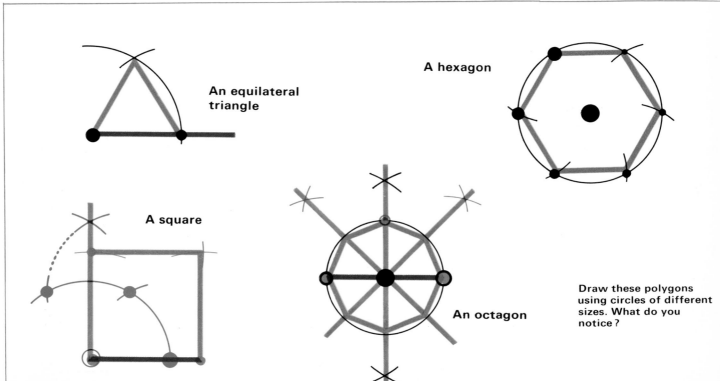

Draw these polygons using circles of different sizes. What do you notice?

Answers pull-out

Exercise 1
a. Shapes which can be made on the square lattice geoboard are those which are numbered 1, 3, 5, 7, 8, 9, 10, 11, 12, 14, 15, 20 and 21.
b. Shapes which can be made on the triangular lattice geoboard are those which are numbered 2, 4, 6, 13, 16, 17, 18, 19, 22, 23 and 24.

Exercise 2
1. The names of the lattice points are given by pairs of numbers. These represent lines of the lattice counted from the position of the red dot.
(2,1) is two lines to the right and one line up.
(−1,3) is one line to the left and three lines up.
(−3,−3) is three lines to the left and three lines down.
(2,−3) is two lines to the right and three lines down.
2.

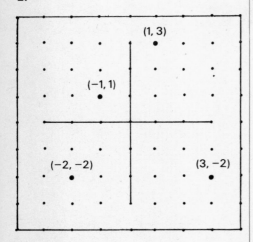

3. (0,0).
4. The number pairs also represent lines of the lattice counted from the position of the red dot.
(2,2) is two lines to the right and two lines upwards.
(−4,2) is four lines to the left and two lines upwards.
(−1,−3) is one line to the left and three lines downwards.
(3,−1) is three lines to the right and one line downwards.
5.

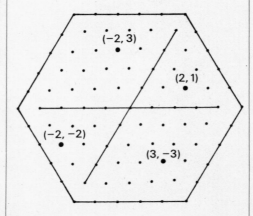

6. (0,0)
7.

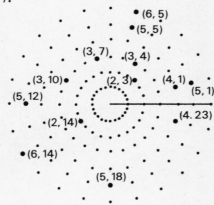

8. The red dot is a fixed point of reference.
9.

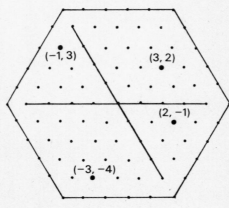

10. (0,0).

Exercise 3
1a. In set A all the first numbers of each pair are the same.
b. The pairs in sets A and B have the same second numbers.
2a. In set D all the second numbers of each pair are the same.
b. The pairs in sets C and D have the same first numbers.
3. Each of the four pins represented by (2,2), (2,−2), (−1,2) and (−1,−2) is enclosed by a red band and by a yellow band.
4. Each red band encloses four pins. Two of each set lie between the yellow bands.
5. Each yellow band encloses five pins. Three of each set lie between the red bands.
6a. The number of pins between the yellow bands on each lattice line is the same.
b. The number of pins between the red bands on each lattice line is the same.

Exercise 4

Colour	Angle in degrees	Angle as fraction of circle	Kind of angle
green	90°	$\frac{90}{360} = \frac{1}{4}$	right
blue	60°	$\frac{60}{360} = \frac{1}{6}$	acute
yellow	120°	$\frac{120}{360} = \frac{1}{3}$	obtuse
pink	270°	$\frac{270}{360} = \frac{3}{4}$	reflex
purple	300°	$\frac{300}{360} = \frac{5}{6}$	reflex
red	240°	$\frac{240}{360} = \frac{2}{3}$	reflex

Exercise 5
1a. 1.   **b.** 2.   **c.** 2.
2a. 3.   **b.** 8.
c. The number of triangles is equal to the number of boundary dots + twice the number of middle dots − 2.
3a. The answer could be i, ii, or iii.

bi.   **ii.**   **iii.**

c. Yes.
d. $T = B + 2M - 2$.

Exercise 6

1.

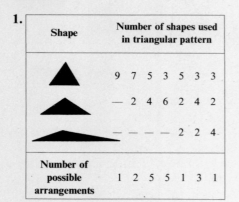

2. The area of triangles in each row are equal being halves of equal wholes. The first and second triangles of row 2 are the same triangles as the second and first of row 1. The third and fourth of row 4 are the same triangles as the fourth and third of row 3.

Exercise 7

1.

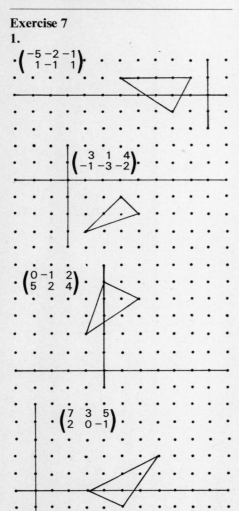

2.

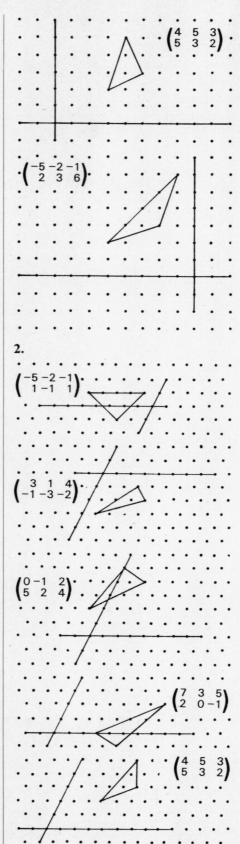

3.

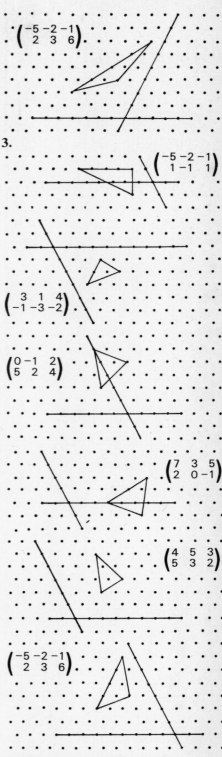

4. The areas of the triangles are, respectively, 8, 4, 7, 8, 5, and 8 three-dot triangles in all three cases. Note that the area of a three-dot triangle on the square lattice is not the same as on the triangular lattice.

Exercise 8

1.

Area of triangle	Ratio	Side	Ratio
$7 = 7 \times 1$	1	$\sqrt{7}$	1
$28 = 7 \times 4$	4	$\sqrt{28}$	2
$63 = 7 \times 9$	9	$\sqrt{63}$	3

2.

Area of square	Ratio	Side	Ratio
$8 = 8 \times 1$	1	$\sqrt{4}$	1
$32 = 8 \times 4$	4	$\sqrt{16}$	2
$72 = 8 \times 9$	9	$\sqrt{36}$	3

3. The areas of similar figures are proportional to the squares of corresponding sides.

Exercise 9

1 and **2.** Compare your measurement with the lattice paper to these values from a table of square roots.
$\sqrt{3}$ is 1·732; $\sqrt{7}$ is 2·646;
$\sqrt{12}$ is 3·464; $\sqrt{28}$ is 5·292;
$\sqrt{63}$ is 7·937.

3. On a triangle lattice the length of the side of an equilateral triangle is the square root of the number of three-dot triangles.

4. On a square lattice the length of the side of a square is the square root of half the number of three-dot triangles.

Exercise 10
1a. $3 + 16 = 19$. **b.** $7 + 21 = 28$.
c. $4 + 9 = 13$.
2a. equal to. **b.** less than.
c. greater than.
3a. $2 + 32 = 34$. **b.** $16 + 26 = 42$.
c. $10 + 16 = 26$.
4a. equal to. **b.** less than.
c. greater than.
5a. A. **b.** O. **c.** R.
6.

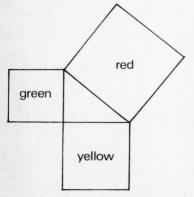

The area, in three-dot triangles, of the red square is 100, of the yellow square is 64, and of the green square is 36. Area of the red square = area of the yellow square + area of the green square.

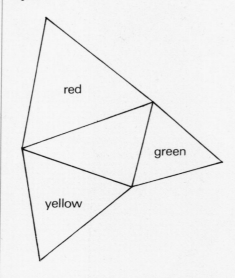

The area, in three-dot triangles, of the red triangle is 25, of the yellow triangle is 16, and of the green triangle is 9. Area of the red triangle = area of the yellow triangle + area of the green triangle.

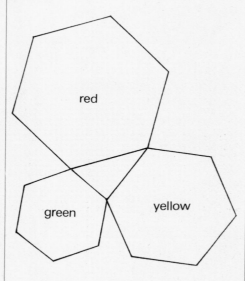

The areas, in three-dot triangles, of the red hexagon is 150, of the yellow hexagon is 86, and of the green hexagon is 54. Area of the red hexagon = area of the yellow hexagon + area of the green hexagon.

Exercise 11

	Quadrilateral												
	1	2	3	4	5	6	7	8	9	10	11	12	13
a.	✓			✓									
b.	✓	✓	✓	✓									
c.												✓	
d.	✓	✓	✓	✓	✓					✓			
e.											✓	✓	
f.					✓	✓	✓	✓			✓		✓
g.	✓	✓	✓										
h.										✓	✓	✓	
i.	✓	✓											
j.					✓	✓							
k.	✓	✓											
l.												✓	
m.					✓		✓	✓	✓				✓
n.								✓	✓				
o.	✓	✓					✓				✓		✓
p.		✓	✓	✓	✓		✓	✓	✓				✓
q.	✓	✓	✓	✓									
r.	✓				✓								

Exercise 12
Sufficient conditions for congruency are given by the following:
2. Side, angle, side (in this order).
4. Side, angle, angle (in this order).
5. Angle, side, angle (in this order).
7. Angle, angle, side (in this order).
8. Side, side, side.
9. Right angle, hypotenuse, side.
(One of the given sides must be the hypotenuse.)
10. Right angle, hypotenuse, side.

Exercise 13
1. Sufficient conditions for congruency of convex quadrilaterals are as follows
Side, angle, side, angle, side

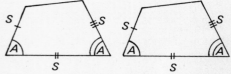

Angle, side, angle, side, angle

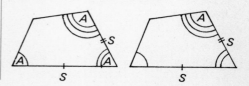

Two sides, and two diagonals at fixed angles to corresponding sides.

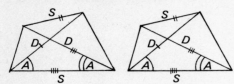

Four sides, and one diagonal joining the vertices of corresponding sides.

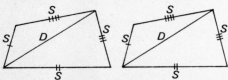

2. For quadrilaterals known to be concave, sufficient conditions are:
Side, angle, side, angle, side

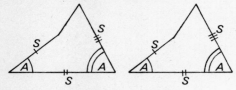

Angle, side, angle, side, angle

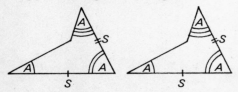

Side, angle, diagonal, angle, side

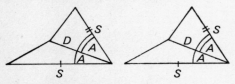

Four sides and one diagonal joining the vertices of corresponding sides.

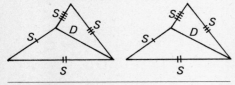

Exercise 14
1. The tangent is perpendicular to the diameter at the point of contact. Angles in a semicircle are right angles.
2. Angles in the same segment of a circle are equal.
3. Purple.
4. The purple angle together with any angle in the minor segment total 180°.

Exercise 15
1. 6 cm.
2. $PQ = 3$ cm, $PR = 2$ cm, $PS = 1.5$ cm.
3. 12 cm.

Exercise 16
1.

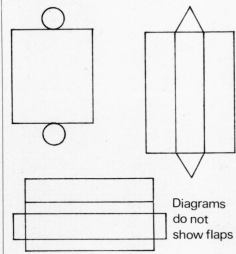

Diagrams do not show flaps

2. Cylinder: A curved surface and two circular faces. When opened out and laid down flat, sides are a rectangle and two circles.
Triangular prism: Three rectangles and two triangles.
Rectangular prism: Six rectangles.
3. In the cylinder the opposite ends are the same. In the triangular prism the opposite ends and the three sides are the same. In the rectangular prism the opposite ends and the opposite sides are the same.
4. Surface area, $A = 2L(B+H)+2BH$.
5. Surface area, $A = 2\pi r(h+r)$.
6. Surface area, $A = 142.8$ cm².

Exercise 17
1. Right-angled isosceles triangles and rectangles.
2. Did you get more than five different patterns? Discuss this exercise with your teacher.

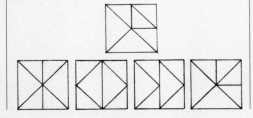

3.

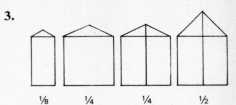

⅛   ¼   ¼   ½

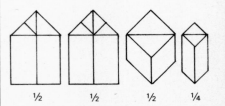

½   ½   ½   ¼

4. ⅓.   **5.** A square
6. ⅙.   **7.** ⅓.
8. ⅛.   **9.** ½.
10. ⅙.   **11.** ½.   **12.** None.
13. The faces of the solid are a combination of regular shapes—a regular hexagon and a square.
14. ½.   **15.** ⅚.
16a. 27.   **b.** 64.

Test
1a. 72°   **b.** 108°.   **c.** 36°.
d. ⅖.   **e.** Ten.
2a. $V = 396$ cm³.   **b.** $V = 132$ cm³.
c. $V = 49.5$ cm³.
d. $\dfrac{Y}{B} = \dfrac{1}{3}$.   **e.** $\dfrac{B}{R} = \dfrac{1}{8}$.   **f.** $\dfrac{Y}{P} = \dfrac{1}{8}$.
3. $XY = 5$ cm.
4. $QR = 10$ cm.
5. The red chord is 8 cm.
6. $R = 75°, B = 105°, P = 133°$.
7a. 90°. (*ABCD* is a *rhombus*. Its diagonals are at right angles to each other.)
b. Corresponding angles.
c. Alternate angles.
d. The line *AC* cuts the pink and green parallel lines, and so angles *QPX* and *XSR* are equal alternate angles. The line *DB* cuts the pink and green parallel lines, and so angles *PQX* and *XRS* are equal alternate angles. Angles *PXQ* and *SXR* are equal. (Vertically opposite angles.)
e. $PX = 3$ cm, $XQ = 4$ cm, $XR = 2$ cm, and $SX = 1.5$ cm.
f. ¼.   **g.** $AC = 12$ cm.   **h.** 96 cm².
8a. $G = 30°$.   **b.** ¼.   **c.** ¼.
d. 30°.   **e.** 120°.

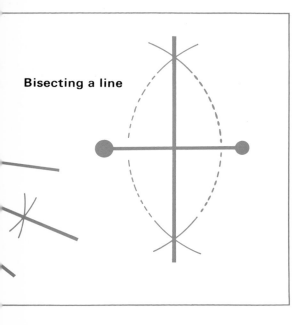

Bisecting a line

A circle outside a triangle

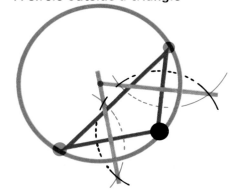

A pentagon

A heptagon

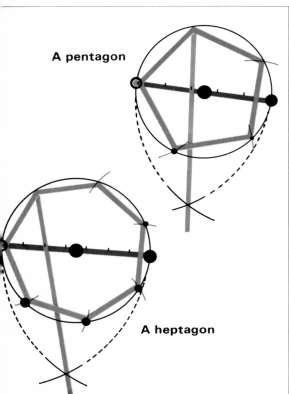

Quadri-laterals

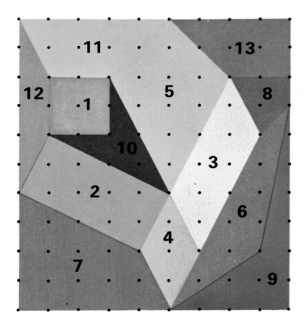

All the shapes in the diagram above have four sides. They are called *quadrilaterals*.

Shapes numbered 11, 12, and 13 are called *trapezia* (singular *trapezium*). Some of the other shapes have special names. Number 1 is a square. Number 2 is a rectangle. Number 3 is a parallelogram. Number 4 is a rhombus. Number 5 is a kite. From the following exercise you should find what special features these shapes have.

Exercise 11
Copy the table below. Use a tick to show which features each quadrilateral has.

Features	Quadrilateral												
	1	2	3	4	5	6	7	8	9	10	11	12	13
a. All sides equal	✓												
b. Opposite sides equal		✓											
c. One pair only of opposite sides equal													
d. Pairs of sides equal			✓										
e. One pair only of sides equal													
f. Unequal sides													
g. Two pairs of parallel sides			✓										
h. One pair only of parallel sides													
i. All angles equal		✓											
j. Opposite angles only equal				✓									
k. All angles right angles		✓											
l. Only two angles right angles													
m. Only one angle a right angle													
n. One angle greater than 180°													
o. Equal diagonals		✓											
p. Unequal diagonals													
q. Diagonals bisect each other			✓										
r. Diagonals bisect at right angles				✓									

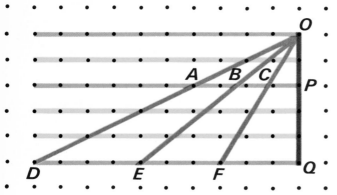

Similarity and ratio

Similarity gives same ratios

Triangles *OAP* and *ODQ* above are similar. Why? *OA* and *OD*, *OP* and *OQ*, *AP* and *DQ* are called *corresponding sides*. *OP* and *OQ* are in the ratio 2 to 5, that is, *OP* is to *OQ* as 2 is to 5. This is often written $OP:OQ :: 2:5$. You will see that $OA:OD :: 2:5$ and that $AP:DQ :: 2:5$.

$$\frac{OA}{OD} = \frac{OP}{OQ} = \frac{AP}{DQ} = \frac{2}{5}$$

Consider triangles *OBC* and *OEF*. *OB* and *OE*, *OC* and *OF* are also in the ratio 2:5.

The diagram below shows a way of checking the ratio for *BC* and *EF*. Take a strip of paper and mark on it the length of *BC*. Move the strip so that *C* is on *E*. Mark *F*. Use an elastic ruler to check that *BF* is divided in the ratio 2:5, that is, $BC:EF :: 2:5$.

Check that the corresponding sides of triangles *OBP* and *OEQ*, *OCP* and *OFQ*, *OAC* and *ODF*, *OAB* and *ODE* are in the same ratio. When triangles are similar, their corresponding sides are proportional.

Look again at triangles *OBC* and *OEF*. On another strip of paper, mark *OB* followed by *BC* followed by *CO*. Mark the four points on an elastic ruler. On the other edge of the strip mark *OE*, *EF*, *FO*. Check with the elastic ruler that $OB:BC:CO :: OE:EF:FO$.

Use the same method with other pairs of similar triangles in the diagram. When two triangles are similar, the sides of one are in the same ratio as the sides of the other.

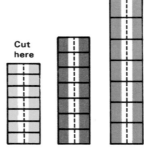

Same ratios give similarity

Cut several strips of paper to different lengths and divide each into seven parts, as shown on page 10. How many different triangles can be made by folding on two marks?

Only two different triangles can be made from seven equal parts. Colour both edges of each strip a distinctive colour as shown. Fold each strip edge to edge and cut it in half. Cut one half on the folds which make one triangle and the other half on those which make the other triangle.

The first diagram below shows three triangles folded in the same way from strips of different lengths. Their sides are in the same ratio because

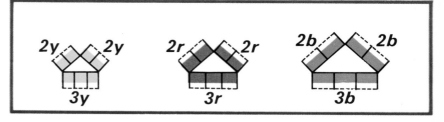

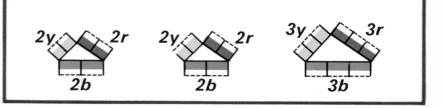

$2y:2b:2r :: 3y:3b:3r :: y:b:r$. Use sectors to check that the triangles are similar. If the sides of a triangle are in the same ratio as those of another, the triangles are similar.

The second diagram above shows three new triangles made from the corresponding sides of the yellow, blue, and red triangles. Their corresponding sides are in the same ratio because $2y:3y :: 2b:3b :: 2r:3r :: 2:3$. Use sectors to check that the triangles are similar. If the corresponding sides of two triangles are in the same ratio, the triangles are similar.

Similarity in quadrilaterals

Pick any four cut lengths and arrange them corner to corner to make a quadrilateral. Change its shape without changing the order of the sides. Do the ratios stay the same? Do the angles stay the same? Draw two quadrilaterals with equal angles but different ratios. Quadrilaterals are similar if and only if they can be divided up into similar triangles.

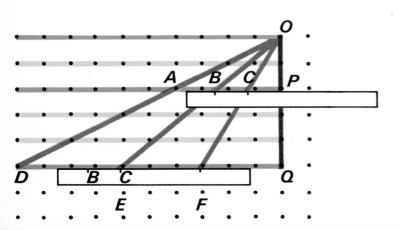

Congruence

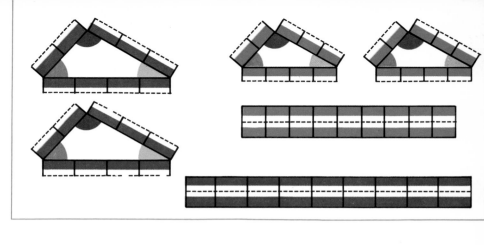

Figures which have the same shape (similar) and the same size (equal area) are *congruent*.

Make four triangles from two nine-fold paper strips of different length, as shown above. Cut out sectors which fit the angles of the green triangles. Test these against the angles of the brown triangles. The four triangles are the same shape but only pairs of the same colour are the same size.

The symbol ≡ means 'congruent to'. If $\triangle ABC \equiv \triangle DEF$, then:

1. $AB = DE$, $AC = DF$, $BC = EF$.
2. $\angle ABC = \angle DEF$, $\angle ACB = \angle DFE$, and $\angle BAC = \angle EDF$.
3. $\triangle ABC$ is equal in area to $\triangle DEF$.

These are *necessary conditions* for congruence. Disproving just one of them would prove that the triangles are not congruent. How many would be sufficient to prove congruence?

In the diagram below, S means side, A angle, H hypotenuse, and R right angle. Equal sides and equal angles are shown in the same colour and by the same letters for each pair of triangles. Which pairs of triangles give sufficient conditions for congruency? For more about congruency see 'Trigonometry' in this series.

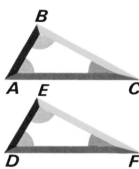

When two triangles are congruent the sides and angles of one equal the sides and angles of the other

Exercise 12

Construct the following triangles. Which sets of information gives sufficient conditions for congruency?
1. 60°, 26°, 94°.
2. 5 cm, 60°, 11·5 cm.
3. 5 cm, 10 cm, 26°.
4. 10 cm, 94°, 60°.
5. 60°, 11·5 cm, 26°.
6. 26°, 10 cm, 5 cm.
7. 26°, 60°, 5 cm.
8. 11·5 cm, 10 cm, 5 cm.
9. 90°, 10 cm, 8 cm.
10. 90°, 10 cm, 6 cm.

Quadrilaterals

Cut four strips of thick card to different lengths. Fasten the ends together to make a quadrilateral. Is it rigid, or can its shape vary? Is its area fixed or variable? If congruence for quadrilaterals means the same shape and the same size, give a set of necessary conditions for congruence. What about sufficient conditions?

A polygon is *convex* when none of its interior angles are greater than 180°. By dividing it up into the smallest number of triangles we can get a set of sufficient conditions for each triangle. Does it also give a set of sufficient conditions for the congruence of the polygon? Would any smaller set be sufficient?

Exercise 13

1. Give a set of sufficient conditions for congruence of convex quadrilaterals.
2. Are the conditions the same for concave and convex quadrilaterals?

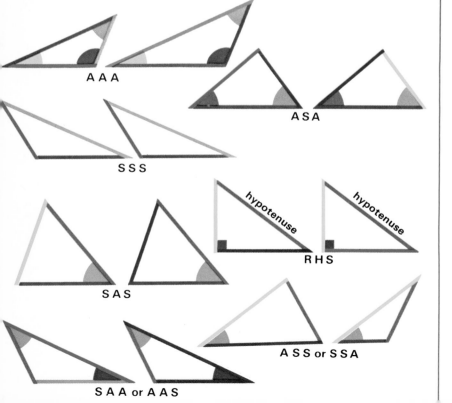

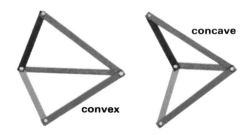

The circle

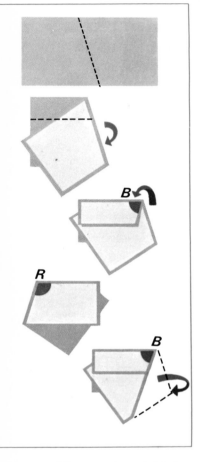

When you have folded the paper and marked the blue angle, turn it over to mark the red angle

Look at the diagram below left. It shows the meaning of some words used on these pages.

Fold a sheet of paper as shown in the diagrams on the left. Call the tip of the blue sector B, and the tip of the red sector R. Take another sheet of paper and pin it to a piece of thick card. In the middle of this sheet of paper, place two pins about 5 centimetres apart.

Place the red sector between the pins, as shown in the first diagram on the right. Keeping both edges of the paper touching the pins, move the sector round. Mark several positions of R with dots.

Turn the folded paper over. You need the angle containing the blue sector. Fold the unwanted part behind so that the fold continues the line of the sector as shown in the last diagram, left. Put the blue sector between the pins from the other side as shown in the second diagram on the right. Move the sector round and mark with dots the path traced by B.

What shape have the dots made? They should look like a circle, but can we assume they are? Here is one way of testing whether the points have made a circle. It consists in doing the experiment in reverse.

Draw a circle on another sheet of paper, roughly in the position shown. Put R on the circumference of the circle and stick pins in where the edges cut the circle.

Making sure that the edges are kept in contact with the pins, move the folded paper and note the path traced by R. Put the blue sector between the pins from the other side. Making sure the edges are kept in contact with the pins, move the folded paper and note the path traced by B.

You should find that R and B move round the circumference of the circle. This shows that the path traced by R and B in the first drawing was also a circle. Sketch in this circular path.

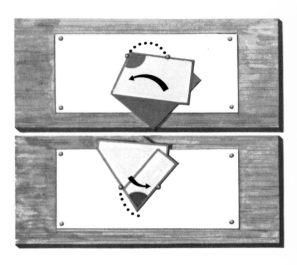

As R moved along the minor arc, the angle marked by the red sector moved in the minor segment. Draw any two positions of the red sector, as shown in the middle diagram below. Such angles are called *angles in the same segment*. They are *subtended* by the minor arc.

Similarly, as B moved along the major arc, the blue sector moved in the major segment. Draw any two positions of the blue sector. These angles are also angles in the same segment. They are subtended by the major arc. Angles in the same segment are equal.

Join the pin-holes in the diagram below right to the centre of the circle. Use the blue sector to find a relation between the size of the angle at the centre and the size of the angle at the circumference. Both angles are subtended by the same minor arc.

The angle at the centre is double the angle at the circumference subtended by the same arc. Use the red sector to check that the same relation holds on the major arc between the angle at the centre of the circle and the angle at the circumference.

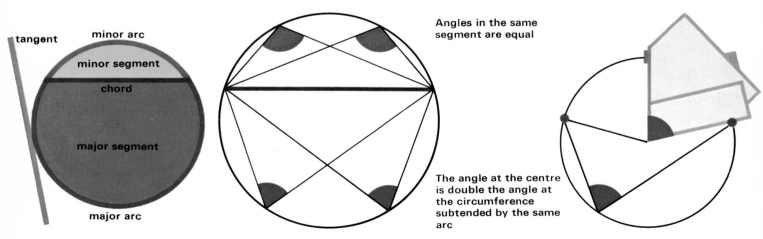

Angles in the same segment are equal

The angle at the centre is double the angle at the circumference subtended by the same arc

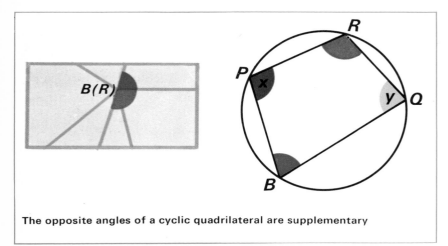

The opposite angles of a cyclic quadrilateral are supplementary

Open out the piece of folded paper. What is the sum in degrees of the red and blue angles? Angles which add up to 180° are called *supplementary angles*.

The diagram above right shows one position of B and one of R. A quadrilateral which can be inscribed in a circle like this is called a *cyclic quadrilateral*. What can you say about the opposite angles B and R? Does this relation hold for the angles x and y? The opposite angles of a cyclic quadrilateral are supplementary.

If you had folded the paper to form right angles, what can you say about the arcs formed? What special name is given to the chord whose end points are represented by the pins in this case? Any angle in a semicircle is a right angle.

In another circle draw a chord. With a piece of paper transfer the chord length to two other positions. For each chord draw an angle in the major segment and one in the minor segment, as shown in the diagram below. Measure the size of the red, turquoise, and green angles. Measure the size of the pink, purple, and yellow angles. What is the sum of the pink and turquoise angles?

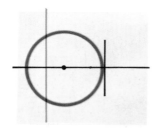

Draw another circle. Fold it along a diameter, making sure that the fold goes through the centre of the pin-hole. Fold a chord perpendicular to the diameter by folding the diameter along itself. Check that the diameter cuts the chord in half. Another way of saying this is the diameter bisects the chord. Fold other chords perpendicular to the diameter. Are they all bisected?

Fold a very small chord. Now make a fold which just touches the circle. This is a *tangent*. It is perpendicular to the radius at the point of contact, as shown in the first diagram, left.

The green lines in the second and third diagrams, left, is the tangent to the circle. A chord meets the tangent at its point of contact with the circle. The angles between the chord and the tangent are shown by the purple sectors. The part of the circle coloured yellow is known as the *alternate segment* in relation to the purple angle. The red angle in each diagram is an angle in the alternate segment.

Exercise 14

The red triangle in the diagram below is made by the diameter of the circle at the point of contact with the tangent, the black chord, and the line joining the diameter to the chord.

1. Why are the angles marked in yellow both right angles?

2. Why are the two angles marked by the grey sectors equal?

3. Since the interior angles of a triangle add up to 180°, the orange angle and the grey angle in the red triangle together make a right angle. What colour should the grey angles actually be?

4. What is the relation between the purple angle between the tangent and chord and any angle in the minor segment?

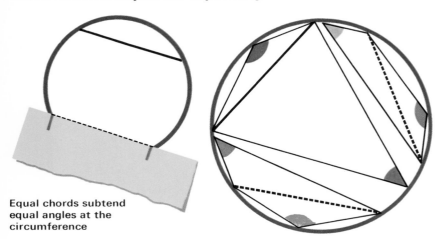

Equal chords subtend equal angles at the circumference

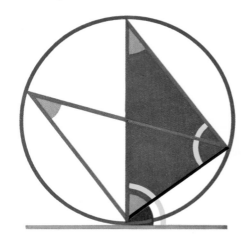

Intersecting chords

A chord and a diameter

Draw and cut out a circle. Fold the circle along a diameter. Open out and mark a point on the diameter. Fold at this point, as shown in the first set of diagrams on the right. Fold again along the first line.

Open out and fold twice again so that the folds join the ends of the diameter to the ends of the chord. The diameter and chord cut each other into two parts called *line segments*. The parts are labelled x, y, and z in the second set of diagrams, right. The line segments of the diameter and chord are coloured. They can be arranged into a square and rectangle with sides equal in length to the line segments. To find the relationship between the area of the square and the area of the rectangle, proceed as follows.

Join the line segments as shown by the pink and brown lines. The green and yellow sectors mark equal angles in the same segment. The triangles formed have the same angles and are, therefore, similar.

The corresponding sides of the triangles are

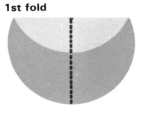
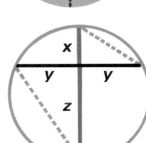

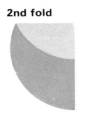
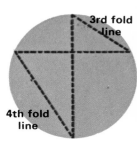
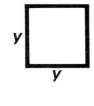
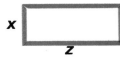
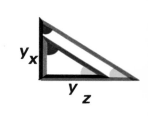

proportional as shown. The ratios of corresponding sides are, therefore, equal.

$$\frac{x}{y} = \frac{y}{z}$$

$$z.x = y^2$$

The area of the rectangle is equal to the area of the black square.

Two chords

Draw and cut out another circle. Fold the circle twice as shown in the first set of diagrams, left. Colour the line segments formed by the intersection. Fold again twice as shown. On the left are two rectangles with sides equal in length to these line segments.

Similar triangles are again formed. Why are their angles equal? The ratios of corresponding sides are equal, giving

$$\frac{a}{c} = \frac{d}{b}$$

$$b.a = c.d$$

The areas of the rectangles formed by the line segments are equal.

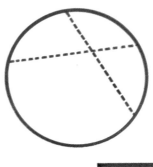

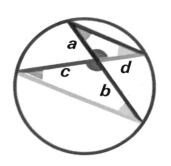
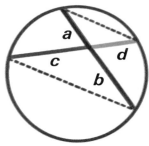
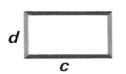
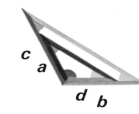

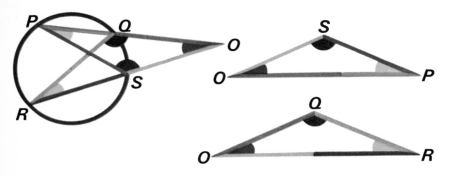
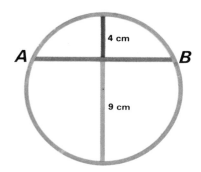

Extended chords

Two chords *PQ* and *RS* are produced, above, so that they intersect at point *O*. Triangles *OPS* and *ORQ* are similar. Why are their angles equal? The ratios of corresponding sides are equal, giving

$$\frac{OP}{OR} = \frac{OS}{OQ}$$

$$OP.OQ = OR.OS$$

Exercise 15
1. Calculate the length of the red chord *AB* intersected at right angles by the diameter.

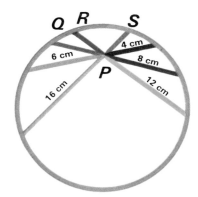

2. Calculate the lengths of *PQ*, *PR*, and *PS*, the purple, blue, and red chord segments.
3. Calculate the length of the orange chord *MN*.

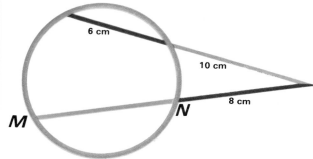

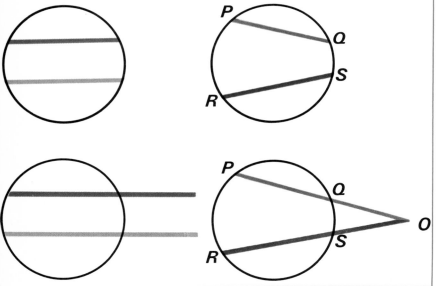

The diagrams above show two more examples of chords which do not meet inside a circle. The blue and pink chords are parallel, and do not meet if extended. If the purple and green chords are extended, they meet outside the circle at point *O*. *PQO* and *RSO* cut the circle at two points. They are called *secants*, *PQ*, *QO*, *RS*, *SO* are called *segments* of secants.

As *OSR* swings down the circle, *R* and *S*, shown on the right by the blue and red dots, get closer together. They coincide at *T* when *OT* is a tangent. The relation *PQ.OQ* = *RO.OS* is then

$$PQ.OQ = OT^2$$

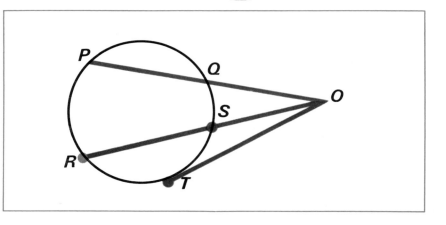

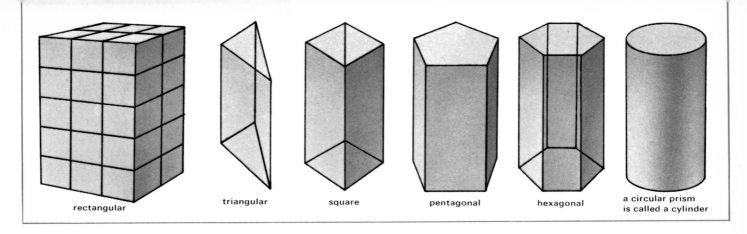

Volume

Prisms

The blue rectangular solid above is made with centimetre cubes from a box of Cuisenaire rods. To calculate the volume of the solid multiply the area of the base, 6 cm², by the height, 5 cm. The volume of the solid is 30 cm³.

This blue solid is called a *rectangular prism*. The other solids above are also prisms. The ends of a prism are polygons of the same shape and size parallel to each other.

The solid on the extreme right of the diagram above is a *cylinder*. It has a circular base and can be regarded as a circular prism. The volume of a cylinder is also calculated by multiplying the area of the base by the height. The area of the base is πr^2, so the volume is given by the formula $\pi r^2 h$.

Pyramids

Pyramids have bases which are polygons, and sides which meet at a single point, called the *vertex*. Two pyramids are shown below.

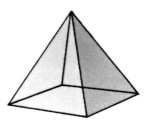 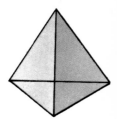

Volume of a pyramid

Make the rectangular box and pyramid shown below from the nets given. Fill the pyramid with sand and empty it into the box. You will need to fill the pyramid three times before the box is full. The volume of a pyramid is one third of the volume of a rectangular solid with the same base and height.

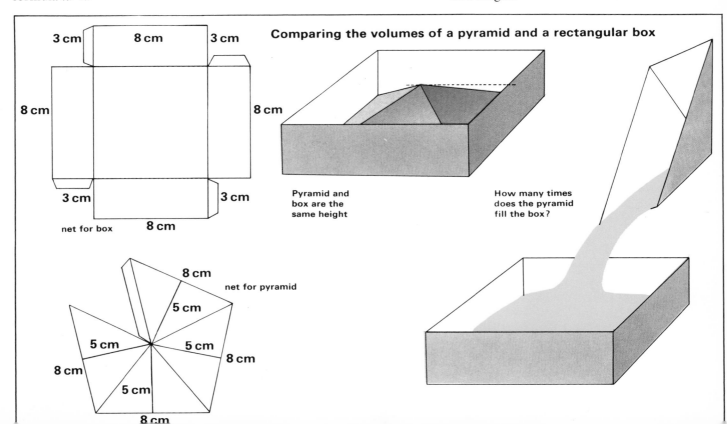

A circular pyramid is called a cone. A triangular pyramid is called a tetrahedron

Comparing the volumes of a cone, a cylinder, and a sphere

Volume of a cone and sphere

You will need a used badminton feather shuttlecock container. Cut off a section the length of the diameter of a tennis ball. Copy the net of the cone below onto thin card. Cut out and glue it together. Check that the cone has the same circumference as the inside of the cylinder and the same vertical height.

Experiment 1
Place the cylinder on a level surface. Fill the cone with sand and empty it into the cylinder. Do this three times. The cylinder should then be exactly full.

The experiment demonstrates that the volume of a cone is one third of the volume of a cylinder which has the same base and height.

Experiment 2
Place the tennis ball inside the cylinder. Fill the cone with sand. Pour the sand into the cylinder until this part of it is full. Empty the cylinder and turn it the other way up. Pour the remaining sand from the cone into this part of the cylinder. The sand should be exactly level with the top of the cylinder.

The diameter of the tennis ball is the same as the diameter and height of the cylinder. It is also the same as the diameter and height of the cone. The volume of the cone is one third the volume of the cylinder. The volume of the tennis ball is, therefore, two thirds of the volume of the cylinder.

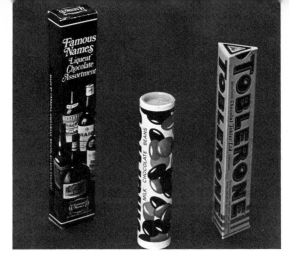

Exercise 16
The above photograph shows three different-shaped containers for sweets and chocolates.
1. Sketch each container opened out flat.
2. Describe the shapes of the sides.
3. Which sides are the same size?
4. The area of a rectangle $= L \times B$. The volume of a rectangular prism $= L \times B \times H$. Write the surface area of the rectangular container in terms of L, B, and H.
5. The circumference of a circle $= 2\pi r$, where r is the radius. The area of a circle $= \pi r^2$. The volume of a cylinder $= \pi r^2 h$. Write the surface area of the cylindrical container.
6. The area of a triangle $= \frac{1}{2}bh$. For the base of the triangular container, b is 3 cm and h is 2·6 cm. The volume of a prism $=$ the area of the base $\times$ the height. The height, H, of this container is 15 cm. Calculate the surface area.

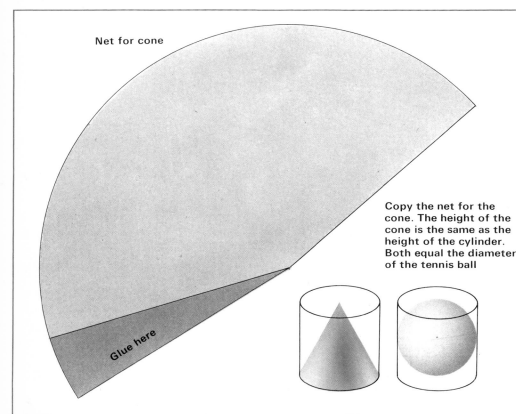

Net for cone

Copy the net for the cone. The height of the cone is the same as the height of the cylinder. Both equal the diameter of the tennis ball

Glue here

Dissections of a cube

The diagrams on the opposite page show how to make eighty models from card. A sketch and net is given for each different model. These models illustrate nine dissections of the cube. The models needed for each dissection are grouped together.

You will need:

A cutting board, a craft knife, a metal straight-edge, a needle mounted in a small handle, a strong quick-drying but thin adhesive, a hard pencil (6H, 7H, or 8H), thin card in a variety of colours.

Make a pinhole template on card for each net using the key shown at the bottom of this page. You will see that you need several copies of most of the nets. Cut them out carefully and glue the joins together. You should find that each set of models assemble together to make the same size of cube.

Notice that each network is drawn in a combination of red, blue, black, and yellow lines. These colours indicate the unit of length of the lines. The diagram below gives the key to the colour code. The sides of the black square are divided into four units. Each black line is a whole number of these units. The red diagonal of the square is divided into four units. Each red line is a whole number of these units. The rest of the diagram shows how the blue and yellow units are constructed.

Key to colour code

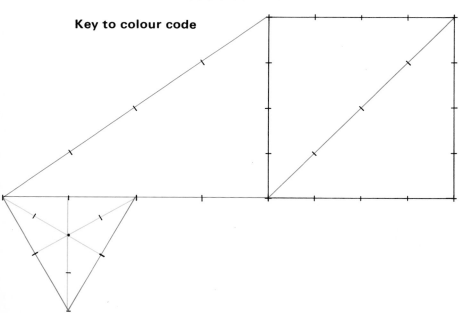

Exercise 17

Cube 1
1. What shapes are the faces of the pieces?
2. How many different patterns can you get on the top face made up of the triangles?
3. Express the various triangular and square prisms formed as fractions of the cube.

Cube 2
4. What fraction is the volume of the pyramid of that of the cube?
5. What shape is a cross-section of the pyramid?

Cube 3
6. The volume of a pyramid is $\frac{1}{3}$ area of the base × the height. A blue pyramid has a base area of half a square. Its height is the same as the cube's. What fraction is its volume of that of the cube?
7. What fraction of the cube is the regular tetrahedron?

Cube 4
8. What fraction of the regular tetrahedron in cube 3 is one of the regular tetrahedrons in cube 4?
9. What fraction of the regular tetrahedron in cube 3 is the regular octahedron (8 faces) in cube 4?
10. What fraction is this regular octahedron of the cube?

Cube 5
11. These pyramids have a face of the cube as base. What fraction is their height of that of the cube?

Cube 6
12. How many regular octahedrons can be made from these pieces?

Cube 7
13. The green pieces in the middle of the cube form a semi-regular polyhedron. Why is this so called?
14. What fraction of the cube is this?

Cube 8
15. The central solid here is another semi-regular polyhedron. What fraction is it of the cube?
16. The eight cubes fit together to make a large cube.
a. How many cubes with edges of 1 cm would be needed to make a cube with edges of 3 cm?
b. How many cubes with edges of 1 cm would be needed to make a cube with edges of 4 cm?

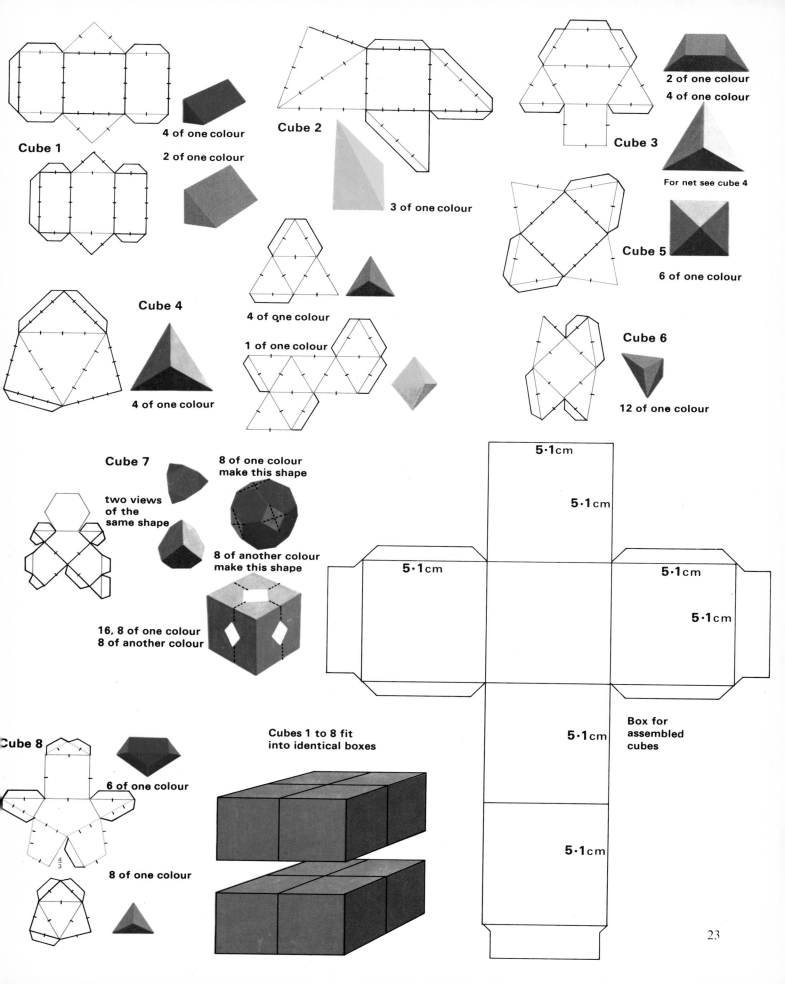

Test

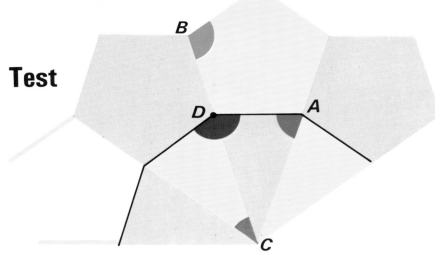
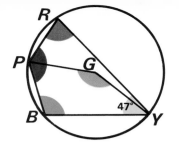

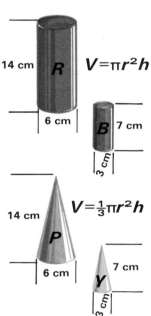

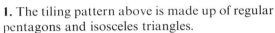

1. The tiling pattern above is made up of regular pentagons and isosceles triangles.
   a. What is the size of the green angle, A?
   b. What is the size of the blue angle, B?
   c. What is the size of the pink angle, C?
   d. What fraction of the space round the black point does the purple angle, D, occupy?
   e. The isosceles triangles form a regular polygon, part of which is shown by the black lines. How many sides has the polygon?

2. Take π to be $\frac{22}{7}$. From the diagrams left, calculate the volume of:
   a. the red cylinder
   b. the pink cone
   c. the brown cylinder.
   Let R, P, B, and Y represent the volumes of the solids. Then $\frac{P}{R} = \frac{1}{3}$. Complete these ratios:
   d. $\frac{Y}{B} =$
   e. $\frac{B}{R} =$
   f. $\frac{Y}{P} =$

3. Calculate the length of the purple chord, XY, in the diagram on the left.

4. The pink line, PT, in the diagram below left is a tangent to the circle at the red point, T. The brown line, PQ, is $\frac{2}{3}$ the length of the pink line. Calculate the length of the blue chord, QR.

5. The red chord, below right, is parallel to the blue diameter. The radius of the circle is 5 cm. The shortest distance from the centre of the circle to the red chord is 3 cm. Calculate the length of the red chord.

6. The green angle occupies $\frac{5}{12}$ of the space round the centre of the circle, G. The yellow angle, Y, measures $47°$. Calculate the size of the red, blue, and purple angles, R, B, and P.

7. $ABCD$ in the diagram below has equal sides. The blue diagonal, DB, is 16 cm. The pink and green lines are parallel to AD.
   a. What is the size of the black angle at X?
   b. What are the yellow angles at Q and R called?
   c. What are the brown angles at P and S called?
   d. Why are triangles PXQ and XRS similar?
   e. $PQ = 5$ cm and $RS = 2\frac{1}{2}$ cm. Calculate PX, XQ, XR, SX.
   f. What fraction of the area of triangle PXQ is the area of triangle XRS?
   g. Calculate the length of the diagonal, AC.
   h. Calculate the area of $ABCD$.

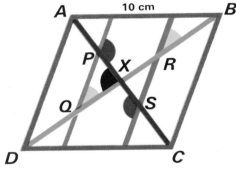

8. The blue line in the diagram below is a diameter of the circle. The purple line is the tangent at O. The yellow angle is $30°$.
   a. What is the size of the green angle at G?
   b. What fraction of the space round point O is occupied by the brown and black angles?
   c. What fraction of the space round point O is occupied by the yellow and brown angles?
   d. What is the size of the pink angle at P?
   e. What is the size of the blue angle at B?

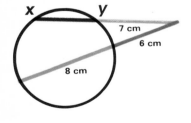
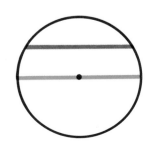
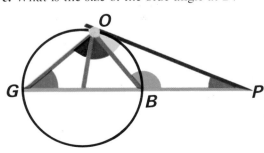